Caribbean Poetry and Paradisiacal Places

DR. RAONA REFIT

Published by R.E.F.I.T PUBLISHING, 2024.

While every precaution has been taken in the preparation of this book, the publisher assumes no responsibility for errors or omissions, or for damages resulting from the use of the information contained herein.

CARIBBEAN POETRY AND PARADISIACAL PLACES

First edition. March 10, 2024.

ISBN: 978-1068655807

Written by DR. RAONA REFIT.

Dedication

This book, a first edition of the REFIT travel book series, is dedicated to *Ms Michelle Ellen Roche Woolf*, a talented and very special Bermudian nursing health and education professional who championed the benefits of travelling away from her own brilliant Bermuda to other beautiful, idyllic, islandic warm locations just like it. She knew how much wellbeing travel could deeply raise one's inner healing, health and wellbeing. She was indeed an early advocate of medical tourism.

As an international clincian, she realised very early in her career that travel for health and wellbeing gains, is an underestated form of powerful and effective wellbing therapy. This fuelled her ambitions and activities as she regularly visited different countries in various continents throughout her career, being increasingly enlightened on the wellness that nature and all theirin provided. Underpinned by her medical knowledge and clinical practice expertise, effortless rapport building and empathetic patient care, she developed awe inspiring abilities to effectively nurse others with kind stories and creative travel sharings. Delivering care to patients, including their assocatied network, she often guided and educated them through their positive rehabiliative journey with visual pictures of advenures encountered and places travelled, evoked through her creative and poetic, travel angel informed words.

Travel for therapy.
Travel for self-care.
Travel to heal.
Travel to share.

CARIBBEAN TRAVEL GUIDE and WRITING INSPIRATIONS CARIBBEAN POETRY and PARADISIACAL PLACES

PREFACE

CARIBBEAN TRAVEL GUIDE and
WRITING PLANNER
CARIBBEAN POETRY and
PARADISIACAL PLACES

By reading this guide you will learn a lot about Caribbean authors and gain a deeper understanding of rich literary creativity of the Caribbean featured through literary works detailing each island. Introducing you to the beauty and history of each island through a sample of written works may foster genuine interest to make deeper connections. Reading this book will provide irresistible value to your itinerary planning, travel writings and checklists as you prepare and enjoy your Caribbean visits.

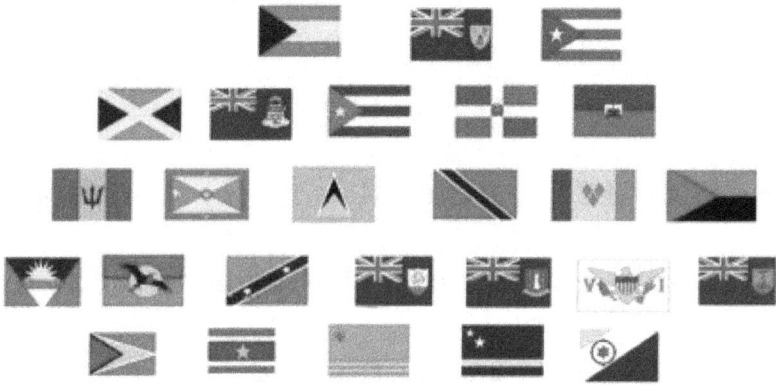

INTRODUCTION

A POETIC TYPE OF TRAVEL GUIDE
PLAN YOUR STAY, PROBE SELECTED
POETS, PREPARE YOUR PARADISIACAL
PLEASURES

Welcome to the R.E.F.I.T. Caribbean travel guide creativity. This exclusive edition includes Caribbean content creations which includes some of the region's most notable literary talent. Be immersed to pages of creativity about different islands that may stimulate you and captivate your interests to forge meaningful connections and visit one (or more) of the islands featured. Be guided on a poetic journey to invigorate interests in learning more and be enlightened on places you may wish to stay at, alongside writers to read about in more detail.

Through these pages, you will tour through a comprehensive journey of the islands through poetry and prose, building an understanding of what may attract you to visit each island. The selection of islands in this edition are from each region of:

- The Greater Antilles
- The Lesser Antilles - Windward Islands
- The Lesser Antilles - Leeward islands and Leeward Antilles
- The Lucayan Archipelago
- The Continental Caribbean

A NOTE FROM THE AUTHOR

DR. RAONA REFIT

You may already know about some information about the ten Caribbean travel destinations featured in this publication.

This book will aim to add to your knowledge and enlighten you further with some impactful literature to complement the beauty of each island.

I will introduce you to a collection of writers famous to each featured island which in turn inspires my own unique travel contributions including information to shape your plans and itineraries that are R.E.F.I.T. Education inspired.

CARIBBEAN ISLANDS SELECTED FOR THIS GUIDE

CREATIVE TRAVEL POETRY IS WRITTEN BY THE AUTHOR ABOUT EACH ISLAND TO GUIDE YOUR TRAVEL PLANS. INSPIRATION HAS BEEN PROVIDED BY A SELECTION OF POETS AND CARIBBEAN AUTHORS AND EXCERPTS OF THEIR WORK.

ANGUILLA
'JEWELS IN SPORT SEA, AND SAND'
Inspired by David Carty – 'Boat racing in Anguilla'.

ANTIGUA AND BARBUDA
'THE TWIN ISLAND – ALL FRUITS RIPE'
Inspired by Joanne C Hillhouse – 'With Grace'

BELIZE
'MAYAN CARIBBEAN CONNECTION'
Inspired by Kalilah (Enríquez) Reynolds – 'The Sun Child'

GRENADA
'MULTICULTURAL TRI-STATE ISLES'
Inspired by Merle Collins – 'Multiculture Abroad'

MARTINIQUE
'MERGING CULTURES AND TRADITIONS'
Inspired by Édouard Glissant – Traité du Toute-Monde

PUERTO RICO
'ESCAPISM THROUGH EXPLORATION'
Inspired by Julia De Burgos – 'The Sea and You'

ST. LUCIA
'NATURAL MYSTICS AND CREATIVE HEALINGS'
Inspired by Kendel Hippolyte – 'Creation'
ST. MARTIN/SINT MAARTEN
'SOUALIGA OR OUALICHI'
Inspired by Lesana Sekou – 'The Great Salt Pond Speaks'
TURKS AND CAICOS
'A LUXURIOUS SEA ROAD FOR SCENIC TRAVEL'
Inspired by Sandra Garland – 'That Kiss on the Forehead'
UNITED STATES VIRGIN ISLANDS
'BEACH AND LAND BRILLIANCE'
Inspired by Tiphanie Yanique – 'The Land of Love and Drowning'

CARIBBEAN ISLANDS AND ASSOCIATED FLAGS

POETIC PERSPECTIVES

The Caribbean is a geographical, historical, and cultural region of the world where a plethora of scholars come from. This book will bring some to name in the land of their beautiful islands which they hail. For those that want to be educated, and for others who want to frame their reasons to travel, this book will invigorate your enthusiasm and curate your artistic creativity. It will aim to leave you with a reason to explore the history of these writers and their Caribbean locations.

POETIC EXCERPTS IN THIS BOOK ARE INSPIRED BY CARIBBEAN POETS

ALISTER HUGHES – CARIBBEAN MAN

Walcott, Louise Bennett, Rhone, Peters, and Hill,
McBernie, Keens-Douglas and many more still,
In drama and poetry, dance, none better than
These greats, they're the soul of Caribbean Man

DEREK WALCOTT

I read.
I travel.
I become.

ISLAND INSPIRATIONS DESTINATIONS & EDUCATION

DR. RAONA REFIT
Education is evolving.
The formulas involved in 'going to school' have changed.
Children become the controllers of the game.
Teachers AI prompt engineer the tasks.
Populations pool, minds are mixing, yet are those within truly learning?
Classes and certificates are no longer counted purely with grades.
Valued papers and perspectives are awarded in virtual places.
Witness-less 'who's' and 'when's' are wistfully taught with weird 'what?!' and 'where?!'
Comments – the new classroom
Machine learning informs the inspectors and advisors.
In this innovation filled AI era – tumbling, twisting, and transforming along on the 4th and 5th revolution rollercoasters.
One-percenters order events
Solipsists strengthen their reserves.
The disconnected off-gridders drive further to disappearance.
Where do you go? What do you want? How do you hope?
The answer may lie in - invigorating your own life itineraries.
Curating your own travel content
Networking amidst new nebulas
Start anywhere, everywhere – just go somewhere! Become a votary of vacations.
Begin your bucket list, develop your dreams, generate your go-to destinations.

Stop supposing. Cut catastrophizing.
Prompt your own path. Engineer your expansion.
Immerse yourself into inspiration intelligence within the creative
Caribbean!

ANGUILLA (Malliouhana)

Island motto: Unity strength and endurance.
BOAT RACING IN AGUILLA
David Carty

It has never ceased to impress me how unique boat racing is in Anguilla.
It may be just another sport to some but to many, it is a festive event and very Anguillan.
It is our own salt watered version of the sport of Kings.

TRAVEL GUIDE POETRY:
'Jewels in sport, sea, and sand'

Natural jewels in Anguilla's crown valley
Malliouhana's Sandy Ground, South Hills, Meads, or Crocus Bay
Form unfettered paradisiacal backgrounds for Anguilla's yearly boat race.
Not just a trending sport for national day, it is their very heartbeat.
Notwithstanding, the Arch location, to behold for hikers, beachcombers, power walkers.
Anyone who wants to experience the island with their feet.
Belmond's Cap Juluca describes its place 'where island bliss finds its meaning'.
Hodge's copper hole invokes that once upon a time never sounded sweeter.
'Moonsplash' King of the Dune Preserve 'Bankie Banks' brings melodic fans every year.
Rendezvous at gorgeous bay beach club Savi
As Carty said it best t'aint no aya – is only we
Anguilla; a beautiful place to be.

ANTIGUA AND BARBUDA
(Wadadli and Wa'omoni)

Island motto: Each endeavouring all achieving.
WITH GRACE
Joanne C Hillhouse

Unlike the rest of the island, the mangoes at Grace's Peak don't bloom only in the summer. Their tantalizing fragrance is a year-round perfume, the trees budding at Grace's will, the juicy fruit plopping into her open palm as she desires, its orange gold sweetness adorning her breakfast plate each morning.

TRAVEL GUIDE POETRY:
'The twin island – all fruits ripe'
A natural sugar destination all fruits ripe!
An array of agribusiness wares of mangoes and pineapple
Entrepreneurs and agro processors offering fruit filled delicacies for all to buy and sample.
Antigua Piango fest – a food tourist's quest
Also, home to famous Richards, Roberts, and Richardson, cricketers best
A twin-island paradise nation – not forgetting Barbuda. A short time away
Codrington Lagoon, Pink Sand beach, Frigate sanctuary; great ways to spend your days.
A fortified location with sights kept at World Heritage site standards.
'Educational tours covering geography and history with expert guides on the parks and forts.
And dedicates special tournament days for all sailing, fishing, and yachting fans.
Whether it be to complement your romance, arts, wellness, or culture.
Antiguan sunshine is a beautiful location for it all while topping up your tans.

BELIZE (Belix)

Island motto: Sub Umbra Floreo.

THE SUN CHILD

Kalilah (Enriquez) Reynolds

I'm a sun child.
I go where he goes,
let light flow,
throw arteries,
brown skin show,
iridescent.

TRAVEL GUIDE POETRY:
'Mayan Caribbean Connection'

A small country that Haynes penned musically as 'land of the free'.
Rich in Mayan culture, its flag rising sun over the sea.
And protected by the world's second largest barrier reef.
Is a great location for water sports giving a once in a lifetime type of experience.
Lush rainforests for birdwatching,
Mayan Mountains for hiking and trail walking
Enchanting visitors with Mayan mystery ruins
Home to some 400 cayes, a beautiful blue hole sinkhole and ATM cave systems
Sharing the vibrancy of kriol culture and garifuna cuisine
Visiting Belize city gives a South American vibe adding to Caribbean dreams.
Their costumes, concerts, parades and road march celebrations bring that tropical niceness.
Heading over to Toledo will showcase more Belize culture, its arts, and culinary spices.
The eco-friendly focus of this quaint location may well entice you.
To stay a little longer because as a tourist it is very affordable too.

GRENADA (Camahogne/Galibi)

Island motto: Ever conscious of God, we aspire, build, and advance as one people.

MULTICULTURE ABROAD

Merle Collins

They want me to write a poem,
a poem like natives write,
about sand sea and sunshine
and exotica like that
They want me to speak a poem.
A poem like they say West Indian speak,
about rice and peas and carnivals and mango trees
and multicultural things like that.

TRAVEL GUIDE POETRY:
'Multicultural tri-state isles'

Grenada - part of a tri-state nation
Where Petite Martinique and Carriacou sister islands are also stationed
Home of Spice Mas carnival
This tri-state multicultural haven named as the world's first culinary capital.
With vibrant festivals and the relaxed living that is compatible
Seasoned with a past of Malcolm X maternal family heritage,
A plethora of literary giants like Hughes, Annim-Addo, Ross, Collins, Skeete and
Wilkins
A population talented, warm and friendly, stalwart yet relaxed in their living.
Entrepreneurial in providing community eco-tourism.
Because Grenadians have such welcoming creative hearts
Evidenced through their attractive creations.
Like Moliniere's Underwater Sculpture Park
Idyllic landscapes show off a marvellous mountainous topography.
And beaches galore all over to frolic and be free.
From St Georges Grande Anse on one island end to Sauters in St Patrick on the
other
Grenada offers and abundance of beach gems.

MARTINIQUE (Madiana)

Island motto: 'La collectivité au service du pays'
(The community at the service of the country)

TRAITÉ DU TOUT-MONDE
Édouard Glissant

I lay hope in the language of landscapes.
The rims of our forests dissolve into the cultivated grounds that reach into the sands.

TRAVEL GUIDE POETRY:
'Merging cultures and traditions'

With titan scholars like Glissant and Cesaire
That championed negritude and creolite moulds.
Martinique is known for its stunning nature and being a pioneer of rhum agricole.
Rum tours galore, scuba diving and water-based activities are fully encouraged
And as a forest filled island for hikers to travel to, there is almost 50% diverse coverage.
The remainder encapsulates beautiful gardens like Balata, and brilliant white, red, black sand beaches.
With the renowned volcanic Mount Pelee region -breathtaking views for all to see
A landscape rich in creole traditions through its language, music, and food
Guides you on a journey that will brighten all tourists' mood.
A Ti punch cocktail from its distilleries will sweeten your inner 'feel good'.
The plethora of unique seafood dishes will go well during festival.
Be free with gwo ka and zouk, shake off the calories by learning the Beguine.
Or take a tailor-made tour with a gommier or yole which is unique to Martinique.
You can glamp in Ti Paradis
Or enjoy Hotel Diamant to stay in luxury.
Either way, memory making it will certainly be.
Whether travelling alone, in couples, with friends or family.

PUERTO RICO (Borikén)

Island motto: Joannes est nomen eius
THE SEA AND YOU
Julia de Burgos
The stroke of the sea upon my door
is blue sensation between my toes
and your impetus leap through my spirit
is on less blue, an eternal birth.
TRAVEL GUIDE POETRY:
'Escapism through exploration'
Boricua's good people - buena gente
Complements the beauty when you enter
The beautiful Puerto Rico
With its nearby Mona Vieques and Celebra in its archipelago
Coral cays and bioluminescent bays
With a history full of Caribbean literacy
And an uplifting contemporary Borucia spirit
Hosting festivals all year round.
Welcoming USA residents and global tourists wherever they are to be found.
There is something to please everyone.
Attractions and activities no matter how large or small your budget may become.
Visit well known locations such as San Juan, El Yunque, Condado or Rincon
And take time out to see the lesser populated Charco Azul tucked away near
Utuado
Or Gilligan's Island near Fajardo
Escape to adventure.
Explore the natural ecosystem.
Elevate your spirit through its architectural beauty.
Enlightening your understanding through its preservation of natural location and
rich history
Or just go with the slow and sustainable travel flow to paradisical Puerto Rico!

ST. LUCIA (Hewanorra)

Island motto: The land, the people the light

CREATION

Kendel Hippolyte

For days, weeks at a time, I lose whatever it is
which keeps my senses softened to the sentience of the earth
to hillside grass running lightly before a silver wind
or a far slope rippling like a muscled shoulder
or how the gradine, faceted pebbles under me will rasp
as I ease in closer, resting my back
against the rough-skinned body of a gliricidia.

TRAVEL GUIDE POETRY:
'Natural mystics and creative healings'

With its tropical climate and fertile soil
An ideal environment for the growth of native trees
From Moringa to Carombola emerges green stick gliricidia
A trekker's, trail seeker's wildlife holidaying dream awaits.
Walcott described it best for all to see.
A place of light with luminous valleys
It is nature's rich healing place in so many ways.
Whether it be a trip to Diamond Falls or Sulphur Springs
The water and soil – mystical properties emanate healing due to the benefits it
brings.
Lauded and celebrated by many musicians for all to hear
Especially on Pigeon's Island, at St Lucia's jazz festival every year
Stay in Rodney Bay, head over to Soufriere.
Unleash your inner explorer, with the natural adventure attractions that are there.

ST. MARTIN/SINT MAARTEN
(Soualiga/Oualichi)

Island motto: Semper progrediens/The friendly island

THE GREAT SALT POND SPEAKS
Lesana Sekou

Salt stings, purifies and preserves, and in these stories,
the people of St Martin are the salt of the land.
They face struggles and hardships
Which they flavour with love and labor
Eventually their past is reconciled, and they claim their very own promised land

TRAVEL GUIDE POETRY:
'Soualiga or Oualichi'

Soualiga or Oualichi – Amerindians first and foremost
The land of brave and beautiful women
The great salt lake filled land
Sensational seas surround a luscious land, with a melting pot of diverse cultures
An ecotourist paradise to behold with a vast range of land and underwater
creatures.
Preserved and curated a pulsating diverse Caribbean palette of people and places
can be found.
Vibrant and artistic from Phillipsburg to Simpson Bay and up to Anse Marcel
A brilliant binational island with a multicultural perspective to show and tell.
A diverse population multiple language they fluently speak.
Enchanting visitors whichever end of the island they seek.

TURKS AND CAICOS (Caya Hico)

Island motto: Beautiful by nature
THAT KISS ON THE FOREHEAD
Sandra Garland

I vowed that as on this road we travel,
Our special bond will never unravel,
Together this journey will make us strong,
And we will renounce the opponents' wrong.

TRAVEL GUIDE POETRY:

'A luxurious sea road for scenic travel'
A small and dreamy archipelago nation
Over forty islands and cays
Houses the beauty of Turks and Caicos
Provo, Grand Turk, and all the cays in Caicos are protected its
expansive barrier reef
A world's best reef and underwater holiday destination
That underpins its sand filled beach creations.
One of the world's smallest countries but big in luxury
Spoiling everyone through its natural beauty
With safety, security and limits on its visitors
Makes TCI an exclusive must see
Whether it be by paddle board and kayaking
Trails and tours to lagoons and caves
Via exclusive cruises to remote island muses
The sea roads travelled shape a gorgeous journey
Showing its mangroves and marine wildlife with huge biodiversity
Offer a great way to live and be free making Caribbean memorable
discoveries.

UNITED STATES VIRGIN ISLANDS

Island motto: United in pride and hope
THE LAND OF LOVE AND DROWNING
<u>Tiphanie Yanique</u>
Yes, we believe in the beach. We have always believed in the beach.
Beaches are places of baptisms and funerals. Of bacchanal but also of
solitude.
TRAVEL GUIDE POETRY:
'Beach and land brilliance
Beach and land brilliance, sapphire through to emerald gems
Water Island and three large saints adorned - Croix, Thomas John
House a plethora of beaches for all Caribbeans and their visitors to
enjoy alone or carnival as one.
Emerging from volcanoes, the regal remnants become bays, rocky
plains and coves for land to sit on.
Rugged and mountainous with farmland in abundance
Famed for its Buck Island National monuments to show off to those
who get the chance.
And the plethora of pretty coral reefs for avid sea life lovers and
seekers to prance and dance.
Outdoor adventures galore with its petroglyphs and hiking trails
Immerse yourself into its magical world of nature and diversity in
creatures and all that it entails.
Visit Bioluminescent bays in St Croix, tour on calm horseback
saunters or engage in zipline thrills
Maybe scuba and sail along St Thomas waters amongst the diverse
marine life therein
Wake up in Gallows Point Carina or Eco Serendib St John

23

Hideaway at Hull Bay St Thomas
Or casino your wallet contents whilst in Divi Carina St Croix
Enjoying a red grout while you sip and recover with a
thirst-quenching cocktail painkiller.

PLACES TO PROBE

The following locations were featured in the travel poetry for each Caribbean Islands. It is highly recommended that you include them in your travel itinerary plans.

For more information about these selected locations contact Dr. RAONA REFIT

ANGUILLA

Anguilla arch, Sandy ground, Crocus Bay, Belmond cap Juluca, Savi beach club, The Dune preserve: Moonsplash festival.

ANTIGUA AND BARBUDA

Codrington Lagoon, Frigate sanctuary, Forts of Antigua, Bogey Peak, Nelson's Dockyard National Park, Cades Bay: Piango festival

BELIZE

Maya mountains, Great Blue Hole, Belize City, Toledo, Belize ancient ruins.

GRENADA

Carriacou, Petit Martinique, Sauters Bay, Spice Mas, Grande Anse Bay, Molieneres' underwater sculpture park.

MARTINIQUE

Mount Pelée, Ti Paradis, Hotel Diamant, Jardin de Balata, Rhum Agricole distilleries.

PUERTO RICO

Mona Vieques, Charco Azul, Rincon, El Yunque, Condado, Gilligan's Island.

ST. LUCIA

Gros Piton, Diamond Falls, Sulphur Springs, Pigeon's Island, Rodney Bay, Jade Mountain.

ST. MARTIN/SINT MAARTEN

Simpsons Bay, Anse Marcel, The Great Salt Pond, Maho, Mullet bay, Philipsburg Great Bay Beach.

TURKS AND CAICOS

Provo, Grand Turk, Little Water Cay tours, Half Moon Bay, West Caicos Marine National Park.

UNITED STATES VIRGIN ISLANDS

St Croix: Altona Lagoon, Salt River Bay
St Thomas: Hull Bay
St John: Eco Serendib Villa, Gallows Point
Divi Carina, Buck Island National Park

POETS TO PREE AND LITERARY LITANIES

Read through the summary profiles of the literary talents mentioned in this guide that hail from each Caribbean location. They have been carefully chosen because their creativity complements their countries. Learn about where you can go in each country they are affiliated with, to gain further literary lessons. This may inspire you to research more about their written creativity and perhaps provide literary enlightenment for you.

For more information about these selected locations contact Dr. RAONA REFIT

POETS TO PREE

ANGUILLA

DAVID CARTY: A writer, historian and cultural activist, David Carty is an Anguillan pioneer with a multifaceted and notable profile across various fields. He is renowned for his artistic work on showcasing Anguillan skills and talent in boat building, racing and sailing.

ANTIGUA AND BARBUDA

JOANNE C HILLHOUSE: A well-known writer, editor, journalist and notable laureate who champions Antiguan literature. Joanne C Hillhouse inspires young authors through her journalistic and literary works in running a non-profit organization which highlights literary artistic talents across Antigua and Barbuda.

BELIZE

KALILAH (ENRIQUEZ) REYNOLDS: Kalilah is an award-winning journalist, educator, entrepreneur, and poet. Her media company creates financial news and education that is easy to understand. It has an underlying mission to help people create and maintain personal wealth and prosperous economies. In addition, she is radio presenter, reporter and television co-host in Caribbean media and entertainment. As a poet and writer with a special passion for developing Belizean literature, she was also involved in pioneering the development of the Belizean Writers Guild.

GRENADA

MERLE COLLINS: A distinguished Grenadian writer and government researcher during the period of the Grenada revolution, Merle Collins has produced poetry collections since 1985. As a distinguished scholar, she has also produced academic and critical works as a Professor Emeritus on themes and trends in Caribbean Writings.

MARTINIQUE

ÉDOUARD GLISSANT: A social theorist, poet and novelist, Édouard Glissant wrote in depth about Martinique throughout the 20th Century. He left a great legacy with a vast and well-respected repository of creative, theoretical literary and academic publications which served to shape the francophone Caribbean 'Creolite' movement.

PUERTO RICO

JULIA DE BURGOS: One of Puerto Rico's most distinguished poets Julia de Burgos was an early advocate of Puerto Rican independence. She is also seen as a foremother of the contemporary Nuyorican poetry movement.

ST. LUCIA

KENDEL HIPPOLYTE: An acclaimed playwright, director and award-winning poet, Kendel Hippolyte is an advocate for social and environmental justice and championing Caribbean arts and literature. He is a co-founder of the Lighthouse Theatre Company in St. Lucia and has a broad base of poetry written in both traditional forms and free form rap/reggae verse.

ST. MARTIN/SINT MAARTEN

LESANA SEKOU: Considered as a prolific author of Caribbean poetry, Lesana Sekou writes with a multifaceted language style and often incorporates Spanish, Dutch and French creole in his individual poems and written works providing recognition of the diversity of his island. An advocate for the independence of St. Martin, which is evident through his works, he is also a publisher, journalist, essayist and an international award-winning literary figure and cofounder of the St Martin book fair.

TURKS AND CAICOS

SANDRA GARLAND: A visionary in telecommunications, Sandra Garland has emerged as an avid contemporary writer of Turks and Caicos. Award winning, Sandra Garland has received national recognition of her poetry publications through the Turks and Caicos Island (TCI) Writers society.

UNITED STATES VIRGIN ISLANDS

TIPHANIE YANIQUE: An academic scholar of creative writing and literature, contemporary author Tiphanie Yanique has achieved critical acclaim across three genres; fictional novels, short stories, and poetry. Describing her work as a social responsibility, she prepares powerful publications which showcase the beauty of her USVI heritage and talents as a Caribbean and American writer.

LITERARY EVENTS
AND LOCATIONS

ANGUILLA

ANGUILLA LIT FEST: Established in 2011, attracting both book lovers and aspiring writers, Anguilla Lit Fest, is a vibrant literary haven. A four-day celebration of literature and self-expression, tourists and citizens can mingle with renowned authors, attend workshops to hone their craft, and participate in lively panel discussions. Soaking up inspiration from literary greats, sharing passions with fellow bibliophiles, and enjoying "Jollification" - Anguilla's unique term for good times - at book signings and poolside cocktail receptions. This island's intimate setting and relaxed atmosphere will spark creativity and ignite a memorable writing journey.

ANTIGUA AND BARBUDA

WADADLI PEN: Based in Antigua and Barbuda by way of the founder Joanne C Hillhouse, the Wadadli Pen literary organization champions the island nation's budding writers. Founded in 2004, this non-profit literary gem fuels creativity through their flagship Wadadli Pen Challenge, an annual writing competition open to all ages. Beyond the competition, Wadadli Pen fosters a love for literature through workshops, school visits, and community events. It's a vibrant space for aspiring authors to find their voice and celebrate Antigua and Barbuda's rich literary heritage.

BELIZE

BELIZEAN WRITERS' GUILD: Unleash your inner poet and delve into the vibrant literary scene of Belize with the Belizean Writers Guild! Founded in 2014, this welcoming community fosters a platform for aspiring writers to hone their craft and share their unique voices. Led by a passionate committee, the Guild champions Belizean stories in global literature, hosting workshops, events, and even publishing member works. Be immersed in the rich cultural tapestry of Belize, drawing inspiration from its diverse landscapes and captivating people. Through the Guild, you can connect with fellow writers, exchange ideas, and learn from established authors. Let the spirit of Belize ignite your creativity and transform your experiences into captivating poems that capture the heart and soul of this extraordinary nation.

GRENADA

GRENADA WRITERS ASSOCIATON: Aspiring tourists and budding writers visiting Grenada will find a supportive community in the island's long-standing writers association. Founded in the 1970s by a group of passionate Grenadian literati, the association has fostered the talents of local writers for decades. This association has a long history of supporting local authors. Reaching out to them online or through Grenada's Ministry of Culture can help connect you with this vibrant group. Over the years, it has provided a platform for workshops, critiques, and discussions, helping shape Grenada's rich literary landscape. The association also plays a role in promoting Grenadian literature abroad, ensuring the island's unique voice finds its place in the wider world.

MARTINIQUE

FESTIVAL EN PAYS REVE: By visiting in July, you can take in Martinique's captivating Festival en Pays Revé, a literary haven that is inspired by the island's rich cultural heritage. This unique event is a great for bibliophiles and aspiring writers to attend. Participate in workshops led by renowned authors, delve into panel discussions exploring diverse genres, or simply soak in the inspiring atmosphere. Mingle with fellow lyricists, attend readings under Martinique's beauty spots and let the island's vibrant energy spark your creativity. Festival en Pays Reve is more than a celebration – it's a literary immersion program promising to ignite your writing journey.

PUERTO RICO

LIBERIA LABERINTO: Nestled in historic Old San Juan, Librería Laberinto is a well-known literary venue that offers a treasure trove of local and international literature, including Puerto Rican works in translation. Browse shelves filled with novels by giants like Julia de Burgos, the poet known as Puerto Rico's "Joan of Arc," or Esmeralda Santiago, whose memoir "When I Was Puerto Rican" captures the immigrant experience with vivid prose. Attend author readings or get lost in translated works by contemporary writers. Let the island's stories, like Rossario Ferré's masterful exploration of identity in "The House on the Lagoon," to ignite travel adventures.

ST. LUCIA

DEREK WALCOTT LIBRARY: Unveiling the literary genius of Saint Lucia, the Derek Walcott Library, housed within the esteemed Sir Arthur Lewis Community College, is a must-visit for all who love to learn and be creatively inspired Sir Arthur Lewis Community College, named after the island's other Nobel laureate, is a renowned institution for higher learning, offering programs in agriculture, arts, science, and technical fields. But the Derek Walcott Library itself deserves special mention. Here, delve into the works of Derek Walcott, Saint Lucia's Nobel laureate for Literature in 1992. Walcott, a masterful poet and playwright, explored themes of identity, history, and the Caribbean landscape in his evocative works. The library, enriched by his personal collection of literary works, allows visitors to gain a deeper understanding of his artistry and the influences that shaped his voice and the beauty of the island. It's a chance to walk in the literary footsteps of a giant and appreciate the enduring power of his words.

ST. MARTIN/SINT MAARTEN

HOUSE OF NEHESI PUBLISHERS: House of Nehesi Publishers is a literary treasure trove for writers and those who like to learn about the island's literary history. Founded in 1970, Mireille Perrin, it became a cornerstone for Caribbean literature, empowering local voices and fostering a unique literary identity. Browse their shelves overflowing with poetry, novels, and historical accounts, all penned by Caribbean authors. House of Nehesi isn't just a bookstore; it's a cultural gem, offering a glimpse into the island's soul and a chance to support the rich literary heritage of the Caribbean. Add it to your itinerary for a truly immersive St. Martin/Maarten experience.

TURKS AND CAICOS

TCI WRITERS SOCIETY: The Turks and Caicos Islands Writers Society (TCIWS). Established in 2019 by Dr. Barbara Ambrister, this group champions local writers, hosting events that spark the creative spirit. Explore bookstores where their venues such as Unicorn Bookstore or House of Unicorn. Their shelves are likely stocked with works by local authors championed by the TCIWS, offering a unique perspective you won't find anywhere else. These locations inspire visitors as they are invited into a vibrant tapestry woven from storytelling, camaraderie, and the shared passion for creative wordplay.

UNITED STATES VIRGIN ISLANDS

Caribbean Writers Society (CWS): Tourists seeking an immersive cultural experience in USVI should delve into the literary world of the Caribbean Writers Society (CWS). This dynamic organization pulsates with the heartbeat of the island's literary scene. Here, unparalleled access to local authors is given through events like "Unveiling USVI's Hidden Gems," where their diverse voices and captivating stories come alive. Beyond the local scene, the CWS might also connect you with regional literary trends and international organizations, providing a broader perspective on Caribbean literature. Not only will you enrich your understanding of USVI's culture but also furnish your travelogue with unique insights and fresh narratives, visiting key sites to shape your creative writing in a way that could develop literary pieces that stand out from the crowd.

LITANIES OF LANDMARKS AND INSPIRATION

ANGUILLA

'The Lit fest'

Established in sunshine, a haven for the word, Lively minds gather, stories to be heard. Lit Fest ignites a passion, a writer's dream takes flight, Luring bookworms and novices bathed in golden light.

Festival bursts with workshops, a chance to refine your skill, sharing secrets of the craft, a master's pen to fulfil. Exchanging thoughts with kindred spirits, a community takes hold, Sparking new ideas, stories yet untold.

Sun-kissed days and starry nights, inspiration takes its hold, sharing passions with fellow bibliophiles, stories to be sold. Infused with Anguilla's spirit, a joyful, rhythmic beat, "Jollification" fills the air, a literary retreat.

Overflowing with book signings, laughter dances in the breeze, Novel encounters bloom beneath the swaying palm trees. Unveiling characters and worlds, imaginations take flight, Memories etched forever bathed in warm Caribbean light.

Tropical paradise meets the power of the written word, inviting minds to wander, emotions deeply stirred. Nurturing creativity, a writer's spirit takes fire, generating journeys on the page, fuelled by pure desire.

ANTIGUA

'Wadadli Pen'

O Wadadli Pen, a beacon bright you shine, based in Antigua, where stories intertwine. Championing voices, both youthful and bold, A haven for writers, their narratives unfold.

Joanne C. Hillhouse, with vision so grand, In 2004, a pen in every hand. A non-profit gem, a literary delight, Fostering creativity, day and darkest night.
Wadadli Pen Challenge, a yearly ignite, sparking imaginations, taking flight. Open to all ages, a dream to pursue, from budding youngsters to seasoned wordsmiths, too.
Beyond competition, your branches extend, Workshops and visits, knowledge to transcend. Schools echo with stories, a love to inspire, nurturing young minds, setting hearts on fire.
Community events, a vibrant embrace, Aspiring authors find their rightful place. Sharing their voices, loud and ever clear, Celebrating the island's rich heritage, held so dear.
Antigua and Barbuda, your stories take wing, Through Wadadli Pen, new voices they sing. Words paint a landscape, vibrant and true, A literary legacy, forever brand new.
With every workshop, with every pen shared, A love for the written word, lovingly declared. From whispers to shouts, a chorus takes form, Antigua's spirit weathers every storm.
O Wadadli Pen, may your influence grow, Inspiring generations, stories to sow. Let pens dance on paper, imagination unbound, A symphony of voices, forever resound.

BELIZE

'Belizean Writers Guild'

In verdant Belize, where stories take flight, the Writers Guild beckons, a beacon so bright. Unleash inner poet, with passion ignite,
Craft honed and nurtured, bathed in warm light, voices unique, a vibrant display, In verdant Belize, where stories take flight.
Workshops abound, with guidance aright, sharing and learning pave the author's way, the Writers Guild beckons. A beacon so bright.
From jungles profound to coral reef's white, Inspiration whispers, stories to say, in verdant Belize, where stories take flight.

GRENADA
'Grenada Writers Association'

Grenada's whispers, writers embrace.
Echo in stories, a vibrant space,
Nurtured by passion from decades untold.
Association gathers, voices bold,
Discussions ignite, critiques set minds free.
Authors empowered, a literary spree,
Workshops unfold, a haven to find.
Rich tapestries woven, for heart and for mind,
Island's own stories, on shelves take their flight,
Traveling outward, bathed in warm light.
Each verse a journey, a landscape unfurled,
Renowned voices heard throughout the world.
Supporting aspiring, with open arms wide,
Advice freely given, a writer's guide.
Shared dreams take root, beneath skies so blue,
On Grenada's shores, inspiration anew.

MARTINIQUE
'Festival en Pays Reve'

In Martinique's heart, where stories take root, Festival en Pays Rêvé, a vibrant shoot.
Writers convene, their dreams take flight,
Inspired by the island's rich, cultural light.
Under palm trees, voices weave a spell,
Festival en Pays Rêvé, a literary well.
Renowned authors lead workshops with might,
Aspiring writers gather, bathed in island light.
Discussions delve in genres, diverse and bright,
Festival en Pays Rêvé, a creative ignite.

Bibliophiles mingle, a joyful sight,
Sharing their passion beneath the starry night.
Readings echo in beauty spots, a lyrical flight,
Festival en Pays Rêvé, a writer's delight.
Martinique's energy sparks a creative might,
Fuelling imaginations, bathed in tropical light.
More than celebration, an immersion so right,
Festival en Pays Rêvé, a literary rite.
Words flow like rum, under the moon's soft light,
Transforming experiences into stories, taking flight.
This literary journey, a beacon ever bright,
At Festival en Pays Rêvé, writers ignite.
In Martinique's embrace, where dreams take flight,
Festival en Pays Rêvé, a festival of light.

PUERTO RICO
'Liberia Laberinto'

Push open the weathered door of Librería Laberinto, and step into a world spun from words. The air hums with a quiet magic, the scent of aged paper mingling with the island's sunshine that spills through arched windows. Here, amongst towering shelves that seem to stretch towards infinity, stories from Puerto Rico and beyond await.

Giant novels stand shoulder-to-shoulder with slim volumes of poetry, their spines whispering promises of adventure, heartbreak, and laughter. Seek out the works of Julia de Burgos, Puerto Rico's Joan of Arc, whose verses blaze with fire and passion. Let Esmeralda Santiago's "When I Was Puerto Rican" transport you to the heart of the immigrant experience, painted in vivid prose. Or perhaps you yearn for a voyage of a different kind? Dive into the masterful "The House on the Lagoon" by Rosario Ferré and explore the shifting landscapes of identity alongside its captivating characters.

Librería Laberinto isn't just a bookstore, it's a portal. Here, local authors weave tales of island life, their voices echoing with the rhythmic pulse of salsa and the whispering palms. Listen as international works, translated with care, paint pictures of faraway lands – each page an invitation to broaden your horizons. Maybe you'll stumble upon a hidden gem by a contemporary writer, their words sparking a conversation or igniting a travel bug within you.

So, wander these labyrinthine aisles, let your fingers trace the worn spines, and lose yourself in the magic of words. For within the walls of Librería Laberinto, stories don't just live on the page – they become your next adventure, waiting to be embarked upon.

.

ST LUCIA
'Derek Walcott Library'.

On Saint Lucia's shores, where emerald hills unfurl, Sir Arthur Lewis College, a beacon for the world.

Named after an island son, an economist so grand, it fosters knowledge, a fertile learning land.

Within its walls, a special space takes hold,

Derek Walcott's Library, a story yet untold.

Another of Saint Lucia's truly treasured, a master of his craft,

His words with wisdom laced a captivating draft.

Unveiling themes of history, a voice of the Caribbean embraced,

He paints a vibrant portrait, a legacy to be traced.

Love and loss intertwine, nature's beauty takes flight.

Exploring identity, bathed in the Caribbean light.

His personal collection, a treasure now on view,

Invites us deeper, where inspiration renews.

First editions gleam, manuscripts whisper untold tales,

Each page a portal, where Walcott's genius prevails.

Delve into poems that shimmer, like sunlight on the sea,

Plays that ignite the stage, with raw humanity.
Explore the landscapes and identities that paint his every page,
From lush rainforests to sun-drenched beaches, a literary pilgrimage.
Here, Walcott's artistry unfolds, a legacy so bright,
Influencers whisperer, shaping words that take flight.
From Homer's epics to Shakespeare's timeless prose,
He weaves a tapestry where history and present flows.
Walk in the footsteps of a giant, his voice a timeless guide,
The Derek Walcott Library, where literary passion cannot hide.
Let his words ignite your soul, a spark to inspire,
To write your own story, fuelled by creative fire.
More than just a library, a space for dreams to take hold,
A place to learn, to grow, a story yet to be told.
So come, explore this haven, where words come alive,
In the Derek Walcott Library, where the Caribbean spirit thrives.

ST. MARTIN/SINT MAARTEN
'House of Nehesi'

Mireille Perrin, a name to remember well, founded House of Nehesi, a story to tell.

In 1970, a treasure chest it became, For Caribbean voices, a literary flame.

Browses overflow, a bibliophile's delight, Poetry, novels, and history take flight. Local authors fill each inky line, Island's soul whispers, truly divine.

Not just a shop, but a cultural gem, where St. Martin's essence finds its diadem. Support the heritage, stories richly spun, A must-visit for your St. Martin fun!

Mireille Perrin, with a vision so grand, in nineteen-seventy, birthed a literary stand. House of Nehesi, a name that now rings, A Caribbean treasure trove, where inspiration sings.

From Philipsburg's heart, a beacon it shines, shelves overflowing with beautiful lines.

Poetry whispers, novels take flight, History's echoes bathed in warm light.

Local voices, nurtured with care, stories of islands, both vibrant and rare.

Identity blossoms, a unique island flair, House of Nehesi, a literary affair.

More than just pages, a cultural embrace, St. Martin's essence, a captivating space.

Support the heritage, a legacy untold. Caribbean stories, both brave and bold.

So explore at your will, this immersive delight, House of Nehesi, where words take flight

For bookworms and learners, a treasure to hold,

A must for St. Martin, facilitating stories to be told.

TURKS AND CAICOS
'TCI Writers Society'

Turks and Caicos whispers, a haven for those who pen,

Championing voices, where stories begin.

Established in purpose, a community to inspire,

Inviting all islanders to set their hearts on fire.

Writers gather, their talents take flight,

Sharing stories beneath the warm, tropical light.

Events stimulate passions, a writer's dream takes hold,

Society embraces, both young and old.

Unicorn Bookstore, a haven for dreams,

House of Unicorn, where imagination gleams.

Shelves bursting with stories, a local display,

Offering a glimpse of the islands' unique way.

Turks and Caicos whispers, a chorus takes flight,

Championing voices bathed in sunlit light.

Established in purpose, in the year of nineteen-nineteen,

Inspiring islanders, a vibrant, literary scene.
Writers gather, a tapestry richly sewn,
Sharing stories, dreams, and seeds to be sown.
Events ignite a passion, a writer's spirit takes flame,
Society embraces, whispering each author's name.
Explore hidden bookstores, where literary treasures reside,
Unicorn Bookstore, a haven where imaginations confide.
House of Unicorn, with shelves that warmly hold,
Stories penned by islanders, both brave and bold.
Offering a window to the islands' unique soul,
You'll find perspectives you won't find elsewhere, whole.
These locations ignite a spark, a creative delight,
Inspiring visitors to join the literary light.
Comradery blossoms, a bond that takes hold,
Sharing the passion for stories yet untold.
Wordplay dances, a symphony so sweet,
Turks and Caicos Writers Society, a vibrant, literary retreat

UNITED STATES VIRGIN ISLANDS
'Caribbean Writers Society'

In US Virgin Islands, where turquoise waves caress the shore, The Caribbean Writers Society beckons, a literary door.
For curious tourists, a cultural immersion so grand, To delve into stories whispered by the island's sun-kissed sand.
This vibrant organization, a heartbeat of the scene, Connects you with local authors, their voices ever keen.
"Unveiling USVI's Hidden Gems," an event that ignites,
Where diverse voices weave tales bathed in moonlit nights.
Beyond the island's shores, the society extends its hand,
Connecting you with regional trends, a literary wonderland.
International organizations join the vibrant fray,
A broader Caribbean perspective to light your literary way.

Enrich your understanding, a cultural tapestry so bright, Unravel USVI's soul, bathed in stories' warm light.

Your travelogue, a canvas blank, awaits the words to bloom, Fresh narratives taking root, dispelling writing's gloom.

Visit key sites that inspire, from bustling Charlotte Amalie's heart
To Trunk Bay's powdery sands, where creative thoughts take part.

Coral reefs teeming with life, a kaleidoscope of hue,
Fuel your imagination, stories waiting to accrue.

With this writer's society being inspiration and guides, your writing will take flight,
Unique perspectives woven, bathed in the Caribbean light.

No longer just a travelogue, but a literary gem,
Standing out from the crowd, a story meant to stem

From the wellspring of inspiration, this island paradise bestows,
Where the Caribbean Writers Society nurtures words that flow.

So come, explore this haven, where culture and stories meet,
And let your travelogue blossom, a literary feat!

POETIC STIMULATIONS AND REPOSITORIES FOR CURATIONS

The following creations have been inspired by the following referenced databases and repositories that you may find useful:

- Caribbean Literary Heritage
- Oxford Bibliographies of each featured author
- Black Cultural Archives
- Caribbean Studies Association
- University of the West Indies SOE Publications

Unleash your creative potential!

The following poems have been created to invite you to read more about associated Caribbean islands mentioned in this guide.

PERUSING THROUGH POETRY
ANGUILLA

Where Anguilla's arch welcomes the dawn,
Sandy Ground whispers secrets bygone.
Crocus Bay beckons with turquoise delight,
Belmond Cap Juluca, a starlit night.
Savi Beach Club, toes in the sand, Tropical cocktails, a drink in your hand.
The Dune Preserve, where music takes flight,
Moonsplash Festival, celebrating sweet rhythms for musical delight.
Sun-drenched beaches and coral reefs grand,
Luxury escapes on this pristine land. Indulge your senses, let worries take flight,
Anguilla awaits, bathed in golden light.

In Anguilla, a Caribbean paradise where the iconic archway frames breathtaking sunrises. Explore the historic village of Sandy Ground, steeped in local charm, before diving into the turquoise oasis of Crocus Bay. Here, crystal-clear waters and vibrant energy promise an unforgettable beach day. For a luxurious escape, unwind at Belmond Cap Juluca, where pampering and starlit nights go hand-in-hand. Craving a lively atmosphere? Savi Beach Club beckons with toes-in-the-sand relaxation and tropical cocktails. Anguilla's magic extends beyond the beach. The Dune Preserve offers a haven for music lovers, especially during the electrifying Moonsplash Festival. Pristine beaches, vibrant coral reefs teeming with life, and the promise of complete indulgence await. Anguilla isn't just a destination, it's a sensory symphony bathed in golden light, ready to wash your worries away.

ANTIGUA AND BARBUDA

In Antigua's heart, Codrington Lagoon gleams,
A turquoise haven, a boater's sweet dreams. Frigate birds soar, a
magnificent sight, Their puffed red throats a breathtaking delight.

History whispers in forts standing tall, Echoes of battles that once used to sprawl. Bogey Peak beckons, a hike to the sky, Panoramic vistas to endlessly sigh.
Nelson's Dockyard, a sailor's domain, Where history and craftsmanship gracefully reign.
Cades Bay's charm, a crescent embrace, Where the Piango Festival fills the space
The people, the music and their rhythms, a cultural blend, Exploring the country will take you on unforgettable journey one that you wish will never end.
Antigua and Barbuda, a paradise found, Treasures abound on this magical ground.

Antigua and Barbuda offer a kaleidoscope of experiences beyond compare. Immerse yourself in the crystal-clear waters of Codrington Lagoon, a haven for boaters and nature lovers. Witness a spectacle of nature at the Frigate Bird Sanctuary, where these magnificent creatures with their puffed red throats put on a breathtaking aerial display. History buffs will be enthralled by the imposing Forts of Antigua, each one a testament to the island's rich past. Feeling adventurous? Hike up to the peak of Bogey Peak, the island's highest point, and be rewarded with panoramic vistas that will leave you breathless. Step back in time at Nelson's Dockyard National Park, a UNESCO World Heritage Site where restored Georgian-era buildings and historic docked ships whisper tales of bygone eras. End your day with a taste of local culture at Cades Bay, a charming crescent-shaped bay. As the vibrant Piango Festival fills the air with music and dance, immerse yourself in Antigua and Barbuda's infectious rhythm. This island paradise promises a unique blend of adventure, history, and cultural immersion, all bathed in the warm glow of Caribbean sunshine.

BELIZE

Belize beckons, a land of mystique, Where Maya Mountains in emerald peaks speak. The Great Blue Hole, a diver's delight, A sapphire jewel in the sunlight.
Belize City, a vibrant embrace, Creole culture with a rhythmic pace. Toledo's heart, a tapestry bright, Garifuna rhythms ignite the night.
Ancient ruins whisper tales untold, Carved temples, mysteries unfold. Xunantunich, Altun Ha grand, Echoes of empires on sacred land.
So, pack your bags, let wanderlust roam, Adventure awaits in this tropical home. Belize, a treasure for all to explore, Where history, beauty, and wonder encore.

Belize beckons with a captivating blend of adventure, culture, and ancient wonders. Delve into the lush embrace of the Maya Mountains, a trekker's paradise offering breathtaking hikes and hidden waterfalls. For the ultimate underwater adventure, plunge into the Great Blue Hole, a natural wonder where divers can explore the depths of a mesmerizing underwater cave. Immerse yourself in the vibrant energy of Belize City, a cultural melting pot showcasing Creole cuisine and a contagious 'joie de vivre'. Venture south to Toledo, a district pulsating with Garifuna rhythms and rich cultural traditions. No journey to Belize is complete without a visit to its ancient ruins. Explore the grandeur of Xunantunich, a sprawling Mayan city steeped in mystery, or marvel at the imposing pyramids of Altun Ha. Uncover the secrets of these bygone civilizations and feel the weight of history at your fingertips. Belize promises an unforgettable experience, where every corner unveils a new adventure.

GRENADA

Grenada's charm, a Caribbean treasured land
Invites you to soak up its sun rays,
Carriacou's whispers lure you to stay. Petit Martinique, a tranquil delight, Secluded beaches, bathed in golden light.

Sauters Bay beckons, a harbour so grand, where yachts gather, and
stories expand.
Spice Mas Grenada, a vibrant display, Costumes of colour light up
the day.
Grande Anse Bay, a crescent of sand, turquoise waters beckon at
your command.
Dive deep in Molinere's underwater art, where its sculptures are
enchanting you to view its magical heart.
Grenada's magic, a spice-scented dream, Adventure awaits on this
island so keen.
So, pack up your bags, and don't delay, Grenada awaits you, come
and stay!

Grenada's allure is a tapestry woven with stunning landscapes, vibrant
culture, and captivating underwater wonders. Island hop to Carriacou,
where time seems to slow down and pristine beaches beckon for
relaxation. Seclusion awaits on Petit Martinique, a tiny gem perfect for
those seeking a tranquil escape. Back on Grenada's mainland, explore
the picturesque Sauters Bay, a haven for yachties and a treasure trove of
local stories. Immerse yourself in the electrifying energy of Spice Mas,
Grenada's renowned carnival, where elaborate costumes and infectious
rhythms paint the streets with joy. Craving sun-soaked paradise? Look
no further than Grande Anse Bay, a crescent of soft sand lapped by
crystal-clear turquoise waters. For a truly unique experience, dive into
Molinere's Underwater Sculpture Park, an enchanting world where art
meets marine life. Grenada offers something for every traveller,
promising an unforgettable blend of cultural immersion, breathtaking
scenery, and unforgettable adventures. So, set sail for Grenada and
discover your own Caribbean paradise.

MARTINIQUE

Martinique's magic, a Caribbean and its sun filed energy calls,
And attractions like Mount Pelée's summit truly enthrals.
Hike through lush rainforests, vibrant and green,

A volcanic island, a majestic scene.
Ti Paradis whispers, a beachcomber's dream,
White sands shimmer, turquoise waters gleam.
Hotel Diamant, a touch of prestige,
Luxury awaits by the sparkling sea's edge.
Jardin de Balata, a floral surprise, a kaleidoscope of colours that
tantalize.
Exotic blooms burst forth in vibrant display.
A photographer's paradise, a breathtaking ballet.
Rhum Agricole, a spirit so fine,
Distilleries beckon, a taste divine.
Sample the essence, a tropical delight,
Where history and flavour in perfect form unite.
Martinique's allure, a seductive blend.
Adventure, relaxation, a story to lend.

Martinique unveils a captivating blend of adventure, exquisite beauty, and rich cultural heritage. Hike the slopes of Mount Pelée, a slumbering volcano that offers breathtaking panoramas and lush rainforests teeming with life. For pure relaxation, sink your toes into the soft sands of Ti Paradis beach, where turquoise waters and swaying palm trees create a picture-perfect paradise. Indulge in unparalleled luxury at Hotel Diamant, a haven where impeccable service meets breathtaking beachfront views. Embrace a world of vibrant colours at Jardin de Balata, a stunning botanical garden overflowing with exotic blooms. Immerse yourself in Martinique's unique heritage by touring a Rhum Agricole distillery. Sample this small-batch rum, produced from locally grown sugarcane, and discover the island's deep-rooted connection to this flavourful spirit. From volcanic peaks to pristine beaches and exquisite gardens, travelling here will set the scene of an unforgettable experience for every kind of traveller to discover the magic of Martinique.

PUERTO RICO

Puerto Rico's got the moves, salsa on the sand,

Mona Vieques whispers secrets, wild and untamed land.
Charco Azul's a jewel, turquoise waters gleam,
Natural pools sparkling, sunbeams in a dream.
Rincon calls the surfers, waves crashing' loud and bold
A symphony of thunder, a story to be told.
El Yunque's emerald whispers, secrets in the trees
A mystical rainforest swaying in the breeze.
Condado's a city symphony, high-rises touch the sky,
Cocktails flowing' freely, as the hours fly by.
Gilligan's Island beckons, a castaway's delight,
Secluded shores whispering', under the starry night.
Bioluminescent magic, city lights so bright, Puerto Rico's got it all,
Head there to see all its glorious sights.
Pack your bags, let your worries unwind,
This island's got a unique rhythm, with a treasure-trove of beauty for
you to find.

Puerto Rico's calling! Immerse yourself in the electrifying rhythm of Puerto Rico, an island paradise pulsating with adventure, relaxation, and cultural vibrancy. This island throbs with adventure, from the wild, untouched beaches of Mona Vieques (don't forget your swimsuit for the glowing bay at night!) to the crashing waves that make Rincon a surfer's paradise. Want to cool off? Charco Azul's natural pools are like stepping into a postcard. Craving a rainforest escape? El Yunque beckons with hidden waterfalls and mystical hikes. Feeling social? Mingle with the locals in Condado, where high-rises meet trendy bars and delicious eats. Need some serenity? Gilligan's Island awaits, a secluded haven with picture-perfect shores. Whether you're chasing bioluminescent wonders, heart-stopping thrills, or a taste of city life, Puerto Rico has it all. Pack your bags and get ready to discover the magic of this unforgettable island!

ST LUCIA

St Lucia is where luxury meets adventure.

Twin volcanic peaks point towards the bright blue sky,
St. Lucia whispers, "Adventure's nigh!"
Hike Gros Piton's slopes, a challenging quest,
Where emerald forests meet the sun's warm crest.
Reach the summit, gasp at the sight below,
Lush valleys unfurl, turquoise waters flow.
Diamond Falls crash, a cool, refreshing spray,
Wash away the travel, let the stress melt away.
Beneath the cascading veil, take a moment to stand,
Feel the jungle's whisper, nature's helping hand.
Sulphur Springs gurgle, a steaming, bubbling sight,
Mud baths and minerals, a natural delight.
Volcanic whispers rise from the earth's hot core,
Rejuvenate your body, and ask for nothing more.
Pigeon Island whispers tales of battles long past,
Explore the ruins, where history's grip holds fast.
Climb the fort's high walls, cannons pointing to the sea,
Imagine battles fought, for all the world to see.
Rodney Bay beckons, a haven for all boat lovers,
Yachts sway gently, a rhythmic silhouette.
Charter a catamaran, sail on sapphire seas,
Explore hidden coves and feel the ocean breeze.
Jade Mountain's magic, a lover's dream come true,
Open walls embrace, with a million-dollar view.
Gaze upon the Pitons bathed in golden light,
Romance and luxury, a perfect, starry night.
Rainforest wonders or volcanic thrills?
St. Lucia's beauty gives you chills.
Kayak down tranquil rivers, hidden and serene,
Spot exotic birdsong in the vibrant green.
Dive beneath the waves, a kaleidoscope so bright,
Coral reefs teeming with life, a dazzling sight.

Swim with playful dolphins, graceful and free,
Underwater magic, just waiting for thee.
Pack your bags, and let go of all your strife,
This island paradise awaits, bathed in vibrant life.
St. Lucia whispers, "Adventure's call,"
Luxury and thrills to conquer one and all

St. Lucia is an island where adventure and luxury collide, promising an unforgettable vacation experience. Hike the majestic Gros Piton, a volcanic peak that rewards you with panoramic vistas of the entire island. Cool off beneath the cascading waters of Diamond Falls, a hidden gem nestled amidst lush rainforests. Take a dip in the world's only drive-in volcano, Sulphur Springs, where therapeutic mud baths and natural mineral pools await. Explore the rich history of St. Lucia at Pigeon Island National Landmark, where historic ruins whisper tales of bygone battles. Mingle with the yachting crowd in Rodney Bay, a vibrant hub of bars, restaurants, and duty-free shopping. For an unforgettable romantic escape, indulge in the unparalleled luxury of Jade Mountain. This resort boasts open-walled sanctuaries with infinity pools, all designed to seamlessly integrate you with the island's breathtaking beauty. From volcanic wonders to rainforest adventures, historical landmarks, and a touch of opulence, St. Lucia caters to every desire. So, pack your bags and discover the magic that awaits you in this captivating Caribbean paradise.

ST. MARTIN/SINT MAARTEN

St. Martin's magic, a melody of two,
French cafes and Dutch charm, waiting for you.
Simpsons Bay whispers, a harbour so grand,
Yachts bobbing gently, a millionaire's band.
Duty-free bounty, a shopper's delight,
Treasures to find in the morning light.
Anse Marcel's song, a tranquil escape,
Turquoise waters lap at a secluded cape.

Palm trees sway softly, a gentle refrain,
Beach bliss awaits, washing worries away.
The Great Salt Pond, a symphony bright,
Flamingos take flight in a rosy, pink light.
Nature's concerto, a chorus of wings,
A spectacle to see, the beauty it brings.
Maho Beach's rhythm, a heart-pounding beat,
Jets skis roar on the water as you tour the waters on this thrilling retreat.
Enjoy the sights and the island's flair,
Absorb the adrenaline rush, with beautiful moments and memories to share.
Mullet Bay's crescent, a surfer's domain,
Waves crash and curl, a turquoise terrain.
Philipsburg's energy, a vibrant parade,
Street vendors and cafes, a colourful cascade.
Music spills out, a joyous embrace, Spicy aromas fill a lively marketplace.
Great Bay Beach stretches, a never-ending sand filled embrace,
Sunsets ignite, painting the sky with grace.
St. Martin's and Maarten's symphony, a dual scenic melody so sweet,
Adventure and charm, on a sun-kissed retreat.

St. Martin/St. Maarten offers a unique opportunity to experience the best of both worlds: French flair and Dutch indulgence. Immerse yourself in the luxurious atmosphere of Simpson Bay, where mega-yachts line the marina and duty-free shops beckon with treasures. Craving tranquillity? Escape to Anse Marcel, a secluded paradise boasting calm turquoise waters and pristine shores. Stroll along the Great Salt Pond, a haven for birdwatchers, and witness the breathtaking spectacle of pink flamingos taking flight. Feel the adrenaline rush at Maho Beach, where low-flying aircraft land mere feet from the shore! For water enthusiasts, Mullet Bay's crescent beach provides world-class waves perfect for surfing. Mingle with the locals in Philipsburg, the bustling capital, where vibrant street vendors and charming cafes line the lively Great Bay Beach. This beach stretches for miles, offering powdery soft sand and breathtaking sunsets that paint the sky in fiery hues. From duty-free havens to heart-stopping plane spotting, St. Martin/St. Maarten caters to every desire. So, pack your bags and discover the magic of this bi-national island paradise!

TURKS AND CAICOS

Turks & Caicos Island travel inspirations can stimulate the words of a beautiful song. Sunshine symphonies where your holiday plans will never go wrong.
Provo's a haven, Grace Bay's a dream. Powdery sands and cocktails by the ocean's gleam. Luxury beckons, restaurants on fire. Indulge and unwind, to lift your good vibes higher
Grand Turk's got history, Cockburn Town's tale. Colonial whispers on a salty exhale. Lighthouse stands tall, a gorgeous setting for enjoying romantic nights.
Or plan for a water based adventure in the early hours of daylight. Dive into shipwrecks, secrets hidden from sight.
Little Water Cay's a nature parade, iguanas chilling in the sunshine's serenade. Kayak through mangroves, a tranquil escape, Nature's wonders, a vibrant scape.
Half Moon Bay's magic, a secluded hideaway. Crystal-clear waters for a perfect getaway. Snorkel with fishes, a colourful spree. Unwind and recharge, let your worries take flight, so your spirit can be free.
West Caicos Marine Park, a diver's delight. Underwater walls, a magnificent sight. Explore hidden caverns, a mystical maze.
Coral gardens dazzling in sun-dappled rays.
Turks & Caicos, a treasure untold. Sun-kissed beaches and stories to unfold.
Island rhythm's calling, can't you hear the beat? Unwind, reconnect, and find your island retreat.

Dive into the turquoise heart of paradise in the Turks and Caicos, where pristine beaches meet unforgettable adventures. Provo, the crown jewel, boasts world-famous Grace Bay Beach with its powdery sand and crystal-clear waters. Luxuriate in pampering resorts and savour delectable cuisine at beachfront restaurants. Yearning for a historical adventure? Journey to Grand Turk, where the charming Cockburn Town whispers tales of the past. Explore the iconic Grand Turk Lighthouse, a timeless sentinel, and delve into the mysteries of shipwrecks that hold forgotten treasures. Nature lovers, rejoice! Little Water Cay beckons with its unique ecosystem. Embark on a thrilling tour to witness the fascinating Turks and Caicos rock iguanas basking in the sun. Kayak through lush mangrove forests, a tranquil haven teeming with life. Half Moon Bay offers a secluded paradise with breathtaking scenery. Imagine snorkelling amidst vibrant coral reefs teeming with colourful fish, or simply relaxing on the pristine shores and letting your worries drift away. For the ultimate underwater adventure, head to West Caicos Marine National Park. Dazzling coral reefs teeming with marine life await exploration. Dive into mystical underwater caverns and marvel at the vibrant coral gardens bathed in sunlight. The Turks and Caicos offer a kaleidoscope of experiences, from luxurious relaxation to thrilling exploration. Choosing this location as your unforgettable island escape will provide you with romantic memories, rest and all the relaxation you deserve.

UNITED STATES VIRGIN ISLANDS

The Virgin Islands reveal their secrets amidst the breeze,
Emerald isles adrift on sapphire seas.
St. Croix, a crown jewel, where magic takes hold,
Altona Lagoon ignites, a story untold.
Glowing waters shimmer, a celestial ballet,
Under a starlit sky, worries melt away.
Salt River Bay murmurs, tales of pirates bold.
Mangrove forests whisper, secrets yet unfold.
Imagine hidden treasures, buried deep below!

Echoes of adventure, where history's whispers flow.
St. Thomas beckons, with beaches of pearl,
Hull Bay's golden sands, a sun-kissed world.
Crystal waves lap gently, a rhythm so sweet.
Adventure and relaxation, a perfect retreat.
St. John unfolds, a nature lover's dream,
Eco Serendib whispers, a tranquil stream.
Luxury unfolds in a verdant embrace.
A haven of peace, a serene, hidden space.
Gallows Point whispers of pirates and lore,
Hidden coves and secrets, waiting to explore.
Trek through emerald forests, dappled with sun,
Uncover island magic, an adventure begun.
Divi Carina beckons, a diver's delight.
Underwater wonderlands, bathed in golden light.
Coral castles rise, in colours so grand,
A breathtaking spectacle, at your command.
Buck Island whispers, a snorkeler's dream.
Sea turtles gliding, in a sunlit stream.
Coral gardens sway, a vibrant display,
Memories made here, forever to stay.
The Virgin Islands sing, a Caribbean song,
Island-hopping adventures, during your vacation you will feel like
you belong. Let your spirit unwind,
For in these emerald jewels, it is a paradise you'll find.

The US Virgin Islands offer a captivating blend of bioluminescent wonders, historical intrigue, and pristine natural beauty. On St. Croix, witness a magical phenomenon at Altona Lagoon, where bioluminescent organisms create a breathtaking nighttime spectacle as they illuminate the water with an ethereal glow. Immerse yourself in history at Salt River Bay National Historical Park and explore the remnants of sugar plantations and Danish colonial architecture. St. Thomas beckons with its sun-kissed shores. Relax on the powdery sands of Hull Bay, where crystal-clear waters invite you to swim, snorkel, or simply soak up the Caribbean sunshine. For a luxurious eco-escape on St. John, look no further than Eco Serendib Villa. This secluded haven promises tranquillity amidst lush vegetation, while still offering easy access to island adventures. Travel back in time at Gallows Point, a historic landmark rumoured to be a pirate hangout. Explore hidden coves and imagine tales of buried treasure! Calling all divers! Divi Carina Bay Resort on St. Croix provides access to some of the Caribbean's most spectacular underwater worlds. Explore vibrant coral reefs teeming with marine life at this diver's paradise. No Virgin Islands adventure is complete without a visit to Buck Island National Park. Snorkel alongside playful sea turtles in crystal-clear waters and marvel at the vibrant underwater coral gardens. From bioluminescent bays and historical sites to luxurious eco-villas and underwater adventures, the US Virgin Islands cater to every traveller's desire.

MORE CARIBBEAN AUTHORS TO FUEL YOUR LITERARY CREATIVITY

In addition to the poets featured on previous pages that come from the islands chosen for this publication, there are many others that hail from each country listed. A selection of more cultural talent for you to read, research and be further inspired can be found in these pages.

Patricia Adams:
Born in The Farrington, Anguilla and renowned as Teacher Patsy, her writings were first featured during the Anguillan revolution. She showcases Anguillan lifestyles of former days to inspire future Anguillans and has a literary award named after her that is given to budding new writers in the Anguillan education system.
Written works include: 'Blue Beans' (2016).

Anguilla

Clement Ashley Banks 'Bankie Banx':
The famous 'King of Dune' reggae singer Bankie Banx, Clement is also an accomplished Anguillan poet who songwriting prowess has origins in being a poet and writer during his teenage years. As a strong activist for youth issues, he founded 'project stingray' to provide music and arts education programs to Anguillan citizens
Written works include 'Soothe your soul', 'The Battle's on' (1982) 'Just cool' (2012).

Jamaica Kincaid
Antiguan born author, Elaine Potter Richardson, is a critically acclaimed writer, whose works details evocative and personally intense recollections from her childhood experiences on the island. Her novels have given rise to controversial opinions due to her outlook of viewing her writing reveals as personal therapy that essentially helped her recover from events in her life.

Antigua and Barbuda

Written works include 'Annie John' (1985) 'Lucy' (1990), 'Mr Potter' (2002).

Corinth Morter-Lewis
An educator, author, poet and playwright, Dr. Morter-Lewis is proud of Belize and has been described as a 'Renaissance woman'. Her work displays her strong affiliation to Belize, its people and provides insight into

Belize

Belize of previous generations. Her poems are introduced to young Belizeans to enthuse gifts to students of what Belize alongside reasons to be proud of their country. **Written works include 'Fathers of Belize' (2001) 'Moments in time' (2013).**

Alister Hughes - Grenada
A renowned Caribbean journalist and poet dubbed 'Prince of Journalism' he was the founder, editor and publisher of 'The Grenada Newsletter'. He is known for being a courageous man of integrity who remained steadfast in his pursuit of truth during the Grenadian Revolution. His vision was to reveal Grenada has a happy, prosperous and multi-island population, united by a common, talented and strong identity.

Grenada **Written work includes: 'Eye Witness to History'.**
(Carriacou **David Ambrose – Carriacou**
and Petite Grenadian author and lecturer of communications
Martinique) studies, Mr Ambrose is a celebrated writer who has Carriacouan familial links. He is also the Director of Grenadian Published Authors (GPA), a non-profit organisation formed in 2022 to provide training and workshops on book writing and publishing. The organisation also provides support to Grenadian authors, hosts reading events, expos and book fairs, encouraging citizens of the island to write.
Written work includes: 'White Spice' (2015), 'That Time in Bogles: A Carriacou Tale' (2018).

Aimé Césaire
Aimé Fernand Césaire was a Martinican multilingual
Martinique poet, author and politician, highly regarded as a founding father of the Francophone Négritude movement in Caribbean literature. The movement was integral was to

promote an appreciation of the history of black people from the French colonies of Africa and the Caribbean. Césaire focused on themes related to resisting powers from colonial domination. As a result of his desire to communicate important messages through his art of writing and garnering audiences to his plays, he works is infamous in literature and arts sectors. His poems and plays are sources of study in both politics and theatrical arts.

Written works include 'Discourse on Colonialism' (translated 1972), 'I Laminary' (translated 1982) 'Lost Body' (translated 1986).

Giannina Braschi

Celebrated as a revolutionary, and cutting-edge author of fiction, poetry geopolitical comedy, Dr. Giannnina Braschi is a charismatic contemporary writer who writes in Spanish, English and the duality of 'Spanglish'. As an academic researcher in Hispanic classic, she has received

Puerto Rico numerous awards for significant contributions to dissemination of Spanish and Hispanic culture. Her native city of San Juan honoured her in 2024, for sharing her depictions of Puerto-Rican culture worldwide.

Written works include: '*Asalto al tiempo* - Assault on Time' (1980), 'Empire of Dreams' (1988), 'Yo-Yo Boing' (1998) 'United States of Banana' (2011).

Derek Walcott

Sir Derek Walcott, is arguably one of the Caribbean's most renowned poet and Nobel Laureate whose planning

St Lucia   committee describes his works as 'a poetic ouverte of great luminosity, sustained by a historical vision and the outcome of a multicultural commitment'. With numerous landmarks such as beautiful outdoor park spaces and indoor knowledge library hubs named after him on his beloved island of St. Lucia, his legacy is cemented and taught to both residents who live and are educated there alongside tourists who visit the beauty of the island. His body of work which spanned over six decades, can be describes as not only moving, poetically flawless and beautifully written, but also timeless, giving weight to the power of literature, leaving a legacy which surpasses literary description and decorative illumination.

Written work includes: '25 poems' (1948) 'In a Green Night' (1962), 'The Gulf' (1970) 'The Fortunate Traveller' (1981) 'Omeros' (1990) 'Tiepolo's Hound' (2000), 'White Egrets' (2010).

Drisana Deborah Jack

St Martin

Both a visual artist and poet hailing from the Dutch half of the island, Dr. Jack is an arts academic and modern poet who produces work that aims to transcend boundaries of style and specific genres. She uses paint, video, photography and text to describe personal and cultural history, with a locus around transcultural existence, the effects of colonialism, memory and mythology. Her work has been published and recited worldwide and her art has been exhibited globally in different continents including USA, Caribbean and Europe.

Written works include: 'The Rainy Season' (1997) and 'Skin' (2006).

Emliy Lorline Malcom

Turks and Caicos

Hailing from one of the smallest islands in Turks and Caicos, South Caicos, Ms Malcolm, focuses on reliving her youthful days on the island, sharing her love for culture through revealing unforgettable stories from her own upbringing on the island. In reading her work, she aims for readers to be immersed into the cultural heritage of Turks and Caicos and its past customs and traditions. As an educator, Ms Malcom communicates the importance of cultural heritage and how to share timeless stories that may resonate with people of all ages and global origins

Written work includes: 'Looking back into yesterday' (2023).

Erika Waters

Dr. Erika J Waters, founding editor of 'The Caribbean Writer', an international literary anthology published by the University of the Virgin Islands is known for publishing critical works on Caribbean literature and women's' literature for over three decades.

United States Virgin Islands

Written works include: 'Critical Issues in West Indian Literature' (1984), 'New Writing from the Caribbean' (1994), 'Contemporary Drama of the Caribbean' (2000).

USEFUL TRAVEL ITEMS

Productive plans pre-empt pleasure in paradise!

Use these pages to remind you of the importance of packing important travel items.

CLOTHES

Workout wear

Beach wear/Towels

Swim accessories

Evening clothes

Outerwear clothes

Lounge wear

Formal wear

TOILETRIES

Shampoo and conditioner

Hair straightener/curlers

Hair styling products

Toothbrush/toothpaste

Shaving/hair removal products

Moisturizer/Deodorant

Make up kit/accessories
Eye care/Mouth care

Important Documents

Passport/travel visas
Driver's license
Reservation confirmations
Travel and event tickets
Travel insurance
Destination maps
Guidebook/Planner

Medical/Beauty

Prescription medications
Pain relievers
Hand hygiene products
Sunscreen
Skin cream
First aid kit/Tweezers
Nail kit/manicure/pedicure accessories

Accessories

Belts/Hats
Shoes/Bags
Jewelry/Scarfs
Water bottle/Food containers
Flip flops/Water shoes
Hair clips/ties
Safety locks

Electronics

Laptop
Smartphone/Tablet
Camera, equipment and accessories
Charging cables for all devices
Head/Earphones
Travel adapters/Power banks

FINAL THOUGHTS

THIS EDITION INCLUDES SELECTED
LOCATIONS WITHIN CARIBBEAN REGIONS:
-THE GREATER ANTILLES ISLANDS
-THE WINDWARD LESSER ANTILLES ISLANDS
-THE LEEWARD LESSER ANTILLES ISLANDS
-THE LUCAYAN ARCHIPELAGO
-THE CONTINENTAL CARIBBEAN
I HOPE YOU HAVE ENJOYED READING THIS
POETIC TRAVEL GUIDE FEATURING A
SELECTION OF CAREFULLY CHOSEN
CARIBBEAN ISLANDS AND ASSOCIATED
LITERARY GREATS.
FUTURE EDITIONS WILL REVEAL MORE OF THE
BEAUTY OF THE CARIBBEAN. EXPLORE AND BE
INSPIRED TO DISCOVER MORE!
DR. RAONA REFIT

FOR MORE INFORMATION:
WWW.REFITWITHRAONA.COM[1]

Don't miss out!

Visit the website below and you can sign up to receive emails whenever DR. RAONA REFIT publishes a new book. There's no charge and no obligation.

https://books2read.com/r/B-A-RVOEB-NGWYC

BOOKS 2 READ

Connecting independent readers to independent writers.

About the Author

Dr Raona R.E.F.I.T is a consultant with vast expertise in allied health and education. She has a special interest in international education, Caribbean literature, economic entrepreneurism and is advocate for promoting healthy lifestyles, fitness and travel wellbeing.

Read more at https://refitwithraona.com/.

About the Publisher

R.E.F.I.T. Publishing is an independent distributor of educational non-fiction and creative fiction literature content in print and electronic formats covering a broad range of topics related to health and wellbeing, Caribbean islands, travel, cultural talent, education innovations and technological entrepreneurism.